वायुमहाशास्त्रम्

Vayu Mahashastram

Ancient Sanskrit Treateise on Climate change , Rainfall , Science of Rainfall prediction & Layers of Atmosphere

Vayu Mahashastram

by Sri Kutupananda Nath

I0490959

© **Copyright**

Edition-1st
March 2023
Chaitra Shukla Pratipada Shaka 1945
Sri Kutupananda Nath
Pune
kutupanandanath@gmail.com
https://vedicvimana.com/

Contents

Vayu Mahashastram of Prabhashankara
Sanskrit with English translation by Sri Kutupananda Nath

Appendices

Drona as unit of measurement of Rainfall in Ancient India

Rainfall Measurements and Rainfall in Ancient India

Layers of Atmosphere as per Vimanashastra

Types of Vatas & Vayu layers from Mahabharata's Shanti Parva

वायुमहाशास्त्रम्

प्रथम भागः
वर्षोत्पत्तिः
गणेशं शिवं भास्करं श्रीगुरुं तं
प्रभाकरं संप्रणम्योथकाञ्ज्या
कृतं प्राक्तनाचार्यविस्तीर्णशास्त्रान्
समाकृष्य पद्यानि मोदाय नृणाम् ॥ १ ॥
यस्मात् क्षुब्धप्रकृतिपुरुषाभ्यां महानस्य गर्भे-'
ऽ हंकारोऽभूत् खकशिखिजलोर्व्यस्ततः संहतेव ।
ब्रह्माण्डं यज्जठरगम ही पृष्टनिष्ठाद्विरंचे-
विश्वं शश्वज्जयति परमं ब्रह्म तत्त्वमाद्यम् ॥ २ ॥
पुराणेष्वंकवेदाश्र कथिता वायवो बुधैः ।
तेषु मुख्या नगा बाता व्याप्तं यैः सचराचरम् ॥ ३ ॥
यूवायुवह इह प्रवहस्तदूर्ध्व :
स्याद्वहस्तदनु संवह-संज्ञकश्च ।
अन्यस्ततोऽपि सुवहः परिपूर्वकोऽस्मात्
बाह्यः परावह इमे पवनाः प्रसिद्धाः ॥ ४ ॥
भूमे हिर्दादशयोजनानि भूवायुरत्राम्बुदविद्युदाचम् ।

(२)
विना समीरावरणं जगत्स्थावर जंगमं
सोनारां प्रपद्येत निर्जना धरणी भवेत् ॥ ६ ॥
भूवायुः कथ्यते यस्तु प्राणपुष्टिसमुद्भवः
युक्तोऽन्वकानिलेनात्र स्वल्पैरन्यैश्व वायुभिः ॥ ७ ॥

संकोचप्रमृतिगुणा वर्तन्तेऽस्मिन्स्वभावतः
ऊर्ध्वं स्थात्स्पर्शनोऽधस्तो भूमेर्धनतरो भवेत् ॥ ८ ॥

उदग्दिशं याति यथा यथा नरः
यथा यथा याम्यदिशं च मध्यात्
वायुस्तथा शीतलतां प्रयाति
तीर्यकरैर्हेलिसमुद्भवैश्च ॥ ९ ॥

वृद्धेः क्षयाच्चापि समुष्णताया
भूमौ प्रवातावरणेषु नियम् ।
गतिर्भवेद्या सजवप्रमाणं
पृथग्विधः कथ्यत एव वायुः ॥ १० ॥

गतिभाजः समीरस्य मुख्या भेदास्त्रयः स्मृताः ।
समीरोऽनिशवाही च नियतोऽनियतस्तथा ॥ ११ ॥

कथ्यते प्रथमो बातो व्यापारो गोलयोः क्रमात् ।
स वाति नितरां वातः ईशानाग्नेयकोणतः ॥ १२ ॥

निरक्षे सर्वदा वाति वायुः पूर्वापरः शुभः ।
स्थलकालान्तराद्भूमेर्गतितं स्यदिग्भवेत् ॥ ११ ॥

समुद्रात्त्वंभसां योगाद्रश्मयः प्रवहन्यपः ॥ १४
तनोत्ययनवशात्काले परिवृत्तो दिवाकरः ।
नियच्छति पयो मेघे शुक्शुकैर्गभस्तिभिः ॥ १५ ॥

अभ्रस्थाः प्रपतत्यापो वायुजाः समुदीरिताः ।
सर्वभूतहितार्थाय वायुभूताः समन्ततः ॥ १६ ॥

मूर्तिमज्जलजं बाष्पं वायोर्निष्क्रम्य दृश्यते ।
मेघस्तदद्द्रद्यते पूर्वैः करकाम्बूद्रवस्ततः ॥ १७ ॥

Translation of chapter one

Vayu Mahashastra.
Part One: The Origin of the Seasons

1. Salutations to Ganesh, Shiva, Bhasakara, and Sri Guru, with the light of knowledge, I gather the extensive teachings of ancient scholars and present these verses to delight the people.

2. From the agitation of Prakriti and Purusha, the great Ahamkara was born, separating the elements of earth, water, fire, and air. Then, from the cosmic egg, Brahma emerged, and from his back came the universe. The ultimate truth is the supreme Brahman, the source of all existence.

3. In the ancient texts and sciences, wise scholars have discussed the nature of the wind. Among them, the primary winds are the Nagas, which pervade all living and non-living things.

4. The youthful winds flow here, above them are the vehicular winds, followed by the Samvaha winds. Beyond these, there are other winds known as Suvaha, Paripurvaka, and the external Paraavaha. These winds are well-known.

5. On Earth, the atmosphere extends ten yojanas high, containing clouds, lightning, and water.

Without the cover of wind, the world of moving and non-moving beings would be desolate,
The earth would become uninhabited and barren. || 6 ||

The air that nurtures life and vitality is called Earth's wind,
It is combined with a little of other winds, present in all directions. || 7 ||

By nature, it possesses the qualities of contraction and expansion,
Upwards, it creates a sense of touch, while downwards, it supports the Earth. || 8 ||

As a person moves towards the east,
And likewise towards the west from the middle,
The wind acquires coolness,
Crossing the sun's rays with its helical motion. || 9 ||

With the increase and decrease of Earth's heat,
A regulated flow of wind exists in the Earth's atmosphere.
This motion is measured by its speed and direction,
The wind is thus described in various ways. || 10 ||

There are three main types of wind movement,
The constant wind, the periodic wind, and the irregular wind. || 11
||

The first type, the constant wind, circulates around the Earth's
poles,
It blows perpetually from the northeast and southeast corners. ||
12 ||

Always blowing without interruption, the auspicious wind moves
from east to west,
Due to variations in geography and time, its direction and motion
on Earth differ. || 11 ||

From the ocean, through the union of waters, rays carry away
water vapor. || 14 ||
At the appointed time, the sun, having completed its revolution,
draws up water,
And deposits it in clouds with dense masses of water droplets. ||
15 ||
From the clouds, driven by wind, fall the waters,
For the welfare of all beings, the wind-blown waters spread
everywhere. || 16 ||

Taking the form of water vapor, it emerges from the water and becomes visible in the air.

The clouds hold it there, where it previously condensed from the vapor. || 17 ||

मेघगर्भाधानाध्यायः २.
अनं जगतः प्राणाः प्रावृड्कालस्य चान्नमायतम् ।
यस्मादतः परीक्ष्यः प्रावृड्कालः प्रयत्नेन ॥ १ ॥
तडिद्वारिमेघानिला गर्जनं च
निमित्तानि मुख्यानिगर्भे भवन्ति
हितं वाहितं चोक्तवत्संपरीक्ष्यः
निमित्तैः बुधैर्वात्र गर्भो घृतो नो ॥ २ ॥
सितादौ मार्गशीर्षस्य प्रतिपद्दिवसे तथा ।
पूर्वापादागते चंद्रेगर्भाणां धारणं भवेत् ॥ ३ ॥
नक्षत्रमुपगते गर्भश्चन्द्रे भवेत्स चंद्रवशात् ।
पंचनवतौ दिनशते तत्रैव प्रसवमायाति ॥ ४ ॥
दिवा भवति यो गर्भो रात्रौ स इति पच्यते ।
धूमेबी पक्षे समुद्भूतः कृष्णे पक्षे च वर्षति ॥ ५ ॥

Chapter 2: Formation of Clouds

The life-giving breath of the world is determined by the arrival of rain,

Hence, the arrival of rain must be examined diligently. || 1 ||

Lightning, rain, clouds, wind, and thunder are the main signs of cloud formation.

One must examine these signs for the proper formation and function of clouds. || 2 ||

In the beginning of the cold season, on the first day of the month of Margashirsha,

When the moon is in the early stage, the holding of cloud formations occurs. || 3 ||

When the moon enters a lunar mansion, the formation of clouds is influenced by the moon.

In ninety-five days, the process reaches completion and rain falls. || 4 ||

The clouds that form during the day bring rain at night, and vice versa.

In the smoky half of the lunar month, rain is produced, and in the dark half, it pours. || 5 ||

पौर्णमास्यामयोत्पन्नः सोऽमावास्यां प्रवर्षति ।
अमावास्यां समुद्भूतः पूर्णमास्यां प्रवर्षति ।। ६ ॥
पूर्वसन्ध्यासमुद्भूतः पश्चिमायां प्रवर्षात ।
पश्चिमायां समुद्भूतः पूर्वसन्ध्यां प्रवर्षति ॥ ७ ॥
पूर्वाइने यः समृद्धतः पश्चाद्रात्रौ प्रवर्षति ।
निशायां पश्चिमे यश्च सपूर्वान् प्रसूयते ॥ ८ ॥
दिनार्थे तु समुत्पन्नः सनिशार्द्धे प्रसूयते ।
निशार्दे तु समुत्पन्नः स दिनार्दे प्रसूयते ॥ ९ ॥

मृगशीर्षाद्या गर्भा मंदफलाः पौषशुरुजाताथ
पौषस्य कृष्णपक्षेण निद्दिशेच्छ्रावणस्य सितम् ॥ १० ॥
माघसितोत्था गर्भाः श्रावणकृष्णे प्रसूतिमायान्ति ।
माघस्य कृष्णपक्षेण निर्दिशेद्भाद्रपदशुक्रम् ॥ ११ ॥
फाल्गुन शुक्तसमुत्वा भाद्रपदस्यासिते विनिर्देश्याः ।
तस्यैव कृष्णपक्षोद्भवास्तु ये तेऽश्वयुक्वा क्ले ॥ १२ ॥
चैत्रसितपक्षजाताः कृष्णेऽश्वयुजस्य वारिदा गर्भाः ।
चैत्रासितसंभूताः कार्तिकशुलेऽभिवर्षन्ति ॥ १३ ॥
पूर्वोद्भूत्थाः पश्चादपरोत्थाः प्राग्भवन्ति जीमूताः ।
शेषास्वपि दिवेवं विपर्ययो भवति वायोश्च ॥ १४ ॥
आङ्कादिददशिवशक्रदिग्भवो मारुतो वियद्विमलम् ।
स्निग्धतिबद्दलपारवेषपरिवर्ती हिमममखाक ॥ १५ ॥

Born on the full moon, he rains on the new moon night,
Emerging on the new moon, he pours down on the full moon night. || 6 ||

Born at the eastern twilight, he rains towards the west,
Originating in the west, he showers on the eastern twilight. || 7 ||

He who prospers in the eastern direction, rains later in the night,
And he who emerges in the western night, gives birth in the east. || 8 ||

Born during the day, he is born again in the middle of the night,
Originating at night, he is born during the daytime. || 9 ||

The pregnancies from Mrigashirsha onwards give poor fruits, so do those born in Pausha,

In the dark fortnight of Pausha, designate Shravana's white one. || 10 ||

Pregnancies born in the bright half of Magha give birth in Shravana's dark fortnight,

In the dark fortnight of Magha, point out Bhadrapada's bright one. || 11 ||

Born in the bright half of Phalguna, they are to be identified in the dark half of Bhadrapada,

And those born in its dark half, indeed, are Ashwayuja's bright one. || 12 ||

Conceived in the bright half of Chaitra, pregnancies are rainy in the dark half of Ashwayuja,

Those born in Chaitra's bright half rain on Kartika's spear. || 13 ||

Those born earlier rain later, and those born later rain earlier, clouds form,

In the remaining days, there is a reversal in the wind's direction. || 14 ||

From the beginning of the tenth day, the wind becomes clear, and the sky is spotless,

With moist, bound clouds swirling around, the hailstones come forth. || 15 ||

पृथुबहुलस्निग्धघनं घनसूचीक्षुरक लोहिताभ्रयुतम् ।

काकाण्डमेचकामं विद्विशुद्धेन्दुनक्षत्रम् ॥ १६ ॥
सुरचापमन्त्रगर्जितविद्युत्प्रतिसूर्य का शुभःसंध्या ।
शशिशिवशक्राशास्थाः शान्तरवाः पक्षिमृगसंघाः ॥ १७ ॥
विपुलाः प्रदक्षिणचराः स्निग्धमयूखा ग्रहा निरुपसर्गाः ।
तरवश्च निरुपस्टष्टांकुरा नरचतुष्पदाहृष्टाः ॥ १८ ॥
गर्भाणां पुष्टिकराः सर्वेषामेव योऽत्र तु विशेषः ।
स्वर्तुस्वभावजनितो गर्भाववृद्धयै तमभिधास्ये ॥ १९ ॥
पौषे समार्गशीर्षे सन्ध्यारागो बुदा परिवेषः ।
नार्थ मृगशीर्षे शीतं पौषेऽतिहिमपातः ॥ २० ॥
शीतमभ्रं तथा वायुश्चन्द्रार्कपरिवेषणम् ।
माघे मासि परीक्षेत श्रावणे वृष्टिमादिशेत् ॥ २१ ॥
फाल्गुने चात्र संघातं वृष्टिस्तनितमेव च ।
पुरोवाताथ ये प्रोक्ता मासि भाद्रपदे शुभं ॥ २२ ॥
बहुपुष्पफला वृक्षा वातः शर्करवर्षणः ।
शीतवर्ष तथाभ्राणि चैत्रेणाश्वयुजं वदेत् ॥ २३ ॥
वहन्ति मृदवो वाताः पुरः शीघ्रं प्रदक्षिणाः ।
विशाखे तानि रूपाणि कार्तिके मासि वर्षति ॥ २४ ॥
मुक्तारजत निकाशास्तमालनीलो तलाञ्जनाभासः ।
जलचरसत्वाकारा गर्भेषु घनाः प्रभूतजलाः ॥ २५॥
तीव्रदिवाकरकिरणाभितापिता मंदमारुता जलदाः ।

The earth is abundant, with dense, moist clouds, and lightning strikes like a crimson arrow.

In the midst of darkness shines the pure moon and stars. (16)

The divine bow emits a thunderous roar, as lightning outshines
the sun at the auspicious twilight.
The tranquil voices of birds and animals fill the quarters, with the
moon, Shiva, and Indra watching over. (17)

The planets move in wide orbits, their gentle rays unobstructed.
Trees and sprouts flourish without hindrance, as humans and
quadrupeds rejoice. (18)

The nourisher of embryos, the exceptional one among all,
Is born of the natural course of seasons, nurturing growth, and I
shall describe him. (19)

In the months of Pausha and Margashirsha, the evening colors
surround Mercury.
In the northern regions, cold prevails during Mrigashira, with
intense snowfall in Pausha. (20)

Cold clouds and winds envelop the moon and sun.
In the month of Magha, observe the rainfall in Shravana. (21)

In Phalguna, there are storms and thunder.
The auspicious winds, as mentioned before, blow in the month of
Bhadrapada. (22)

Abundant trees bear flowers and fruits, and the winds shower sweet rain.
In the months of Chaitra and Ashvayuja, speak of cold rain and clouds. (23)

The winds swiftly carry the soil in a clockwise direction.
In the month of Kartika, it rains in the constellation of Vishakha. (24)

The lustrous pearls and silver, the dark blue Tamala, and the black Anjana glow.
Numerous water-dwelling creatures inhabit the dense and abundant waters of the womb. (25)

The clouds, scorched by the intense rays of the sun, are cooled by the gentle breeze. (26)

गर्भधारणाध्यायः ३.
॥
ज्येष्ठस्य शुक्लाष्टम्यां तु नक्षत्रे भगदैवते ।
चत्वारो धारणाप्रोक्ताः मृदुवातसमीरिताः ॥ १
नीलांजननिभैर्मेघैर्विद्युत्स्थागतमारुतैः ।

विस्फुलिंगरजोधूवैरच्छनौशशिदिवाकरौ ॥ २ ॥
एकरूपाः शुभाज्ञेया अशुभाः सान्तराः स्मृताः ।
अनार्यैस्तस्करैर्घोरैः पीडा चैव सरीसृपैः ॥ ३ ॥
ततः स्वात्यादि नक्षत्रैश्चतुर्भिः श्रावणादयः ।
परिपूर्णाः शुभास्ताः स्युः सौभ्याः शिवसुभिक्षकाः ॥ ४ ॥
स्वातौ तु श्रावणं हन्यात् वृष्टिश्वेन्द्राग्निदैवते ।
भाद्रपदे त्ववृष्टिः स्यात् मैत्रे चाश्वयुजे स्मृता ॥ ५ ॥
ऐन्द्रे तु कार्तिकेऽप्येवं वृष्टे वृष्टिं निर्हति च ।
एतेषु यदि नो वृष्टिस्तदा सौभिक्षलक्षणम् ॥ ६ ॥
यदा तु विद्युतः श्रेष्ठाः शुभाशाः प्रत्युपस्थिताः ।
सपांशुवर्षा सापश्च शुभा वालक्रिया अपि ॥ ७ ॥
पक्षिणां सुस्वरा वाचः क्रीडापांशुजलादिषु ।
रविचंद्रपरिवेषाः स्निग्धानात्यन्तदूषिताः ॥ ८ ॥
मेघाः स्निग्धाः संहताच प्रदक्षिणगतिक्रियाः ।
तदास्यान्महती वृष्टिः सर्वसस्याभिवृद्धये ॥ ९ ॥
।
इति वायुचकशास्त्रे गर्भधारणाध्यायस्तृतीयः

Garbhadharana Chapter 3

1. On the eighth day of the waxing moon in the Jyeshtha month,
with the constellation of Bhagadaivata,
Four types of gentle winds are said to be present, accompanied by
soft breezes.

2. With the appearance of dark-blue clouds, filled with intelligence and lightning-born winds,
The sky is adorned with flickering sparks and dust, like the moon and the sun.

3. Those with a uniform appearance are considered auspicious, while those with irregularities are deemed inauspicious,
Causing distress through the actions of uncivilized people, robbers, and venomous creatures.

4. From the Swati constellation onwards, up to the four Shravana constellations,
They become complete, auspicious, and bring prosperity, peace, and good harvests.

5. In the Swati constellation, Shravana is destructive, with the sight of Indra-Agni deities,
In Bhadrapada, there is a lack of rain, and in Ashwayuja, it is known to be friendly.

6. In the Aindra constellation, even during Kartika month, rain is believed to fall and cease,
If there is no rain in these constellations, it indicates a good harvest.

7. When the best lightning flashes and auspicious rainbows appear,
There will be partial rain, and even the appearance of comets is auspicious.

8. The melodious speech of birds playing in the water and other elements,
The enveloping aura of the sun and the moon, neither too oily nor too polluted.

9. Clouds that are smooth, closely gathered, and move in a clockwise direction,
At that time, there will be abundant rain for the growth of all crops.

Thus concludes the third chapter on Garbhadharana in the science of wind and clouds.

वृष्टिशुभाशुभज्ञानाध्यायः ४.
ज्येष्ठे मूलमतिक्रम्य मासि प्रतिप्रदतः
वर्षासु वृष्टिज्ञानार्थं निमित्तानुपलक्षयेत् ॥ १ ॥
ज्येष्ठे मूलमतिक्रम्य पतन्ति विन्दवो यदा
प्रवर्षणं तदाज्ञेयं शुभं वा यदि वाशुभम् ॥ २ ॥

ज्येष्ठ कृष्णे च पूर्वायाम् जेष्टयाश्चैवान्तरेतदा ।
देवः प्रतिपदायान्तु यदाकुर्यात्प्रवर्षणम् ॥ ३ ॥
चतुषटीन्याढकानी तदा वर्षाति माधवः ।
निष्पद्यन्ते च सस्यानि विगतोपद्रवाणि च ॥ ४॥
धर्मार्थकामा वर्तन्ते परचक्रं प्रणश्यति ।
सुभिक्षं क्षेममारोग्यं दशरात्रंत्ववगृदः ॥ ५ ॥
उत्तरायामषाढायां यदा देवः प्रवर्षति ।
विज्ञेया द्वादशद्रोणाः ज्ञेयं वर्ष सुभिक्षदं ॥ ६ ॥
तदा निम्नानि वाप्यानि मध्यमं वर्षणं भवेत् ।
सस्यानां चापि निष्पतिः सुभिक्षं क्षेममेव च ॥ ७ ॥
श्रवणे खारिका ज्ञेया सस्यं च निर्दिशेत् ।
चौराश्च प्रबलाज्ञेया व्याधयश्च पृथग्विधाः ॥ ८ ॥
क्षेत्राण्यत्र नरोहंति न दर्पो नास्ति जीवितं ।
अष्टादशाहं जानीयाद् अपग्राहं न संशयः ॥ ९ ॥
आढकानि घनिष्ठायां सप्तपंचाशदादिशेत् ।
मही सस्यवती ज्ञेया वाणिज्यं च विनश्यति ॥ १० ॥

Chapter 4: Knowledge of Auspicious and Inauspicious Rains

1. After crossing the Mula constellation in the month of Jyeshtha, one should observe the omens for rainfall during the year.

2. When the winds blow after crossing the Mula constellation in Jyeshtha, one should know the commencement of rains, whether it is auspicious or inauspicious.

3. In Krishna Paksha of Jyeshtha, during the early part of the day and when the asterism is Jyeshtha, if the rain god showers on the first day, then:

4. Madhava (Vishnu) sends rain for 64 days; the crops grow, and the troubles cease.

5. Dharma, Artha, and Kama prosper; the enemy's power declines. For ten nights, there will be abundant food, safety, and health.

6. When rain falls during Uttara Ashadha, one should consider twelve dronas of rainfall and a year of abundant food.

7. Then, the low-lying reservoirs get filled, and there is moderate rainfall. The growth of crops, abundant food, and safety also occur.

8. In Shravana, one should know the Kharika and predict the crops. Thefts and various diseases should be considered prevalent.

9. People plow the fields here, but there is no pride in living. One should know that there will be a drought for eighteen days without a doubt.

10. In Ghaniṣṭha, one should predict seventy-five half-dronas of rain. The land will be full of crops, but trade will decline.

क्षेमं सुभिक्षमारोग्यं पंचरात्रमवग्रहः ।
प्रबला दंष्ट्रिणोज्ञेया मूषकाः सलभाः शुकाः ॥ ११ ॥
खारी तु वारुणे ज्ञेया सस्यानां चाप्युपद्रवः ।
चोरास्तु प्रबला ज्ञेया न च कश्चिदपग्रहः ॥ १२ ॥
पूर्वाभाद्रपदायान्तु यदा देवः प्रवर्षति ।
चषीन्याकानि तदा वर्षात माधवः ॥ 13 ॥
सर्व धान्यानि जायन्ते बलवन्ताश्च तस्कराः ।
नायके क्षुभिते चापि दशरात्रमपग्रहः ॥ १४ ॥
आढकानां च नवतियुत्तरायाम् समादिशेत् ।
स्थलेषु वापयेद्विजं सर्वतस्य समृद्ध्यति ॥ १५ ॥
क्षेमं सुभिक्षमारोग्यं विंशरात्रमपग्रहः ।
दिवतानां विजानीयाद्भद्रबाहुवचो यथा ॥ १६ ॥
चतुषष्ट्याढकानीह रेवसामपिनिर्दिशेत् ।
सस्यानि च प्ररोहन्ति सर्वाण्येव न संशयः ॥ १७ ॥
उद्यन्ते च राजानः परस्परविरोधिनः ।
ये च मुख्यास्तु शोभन्ते वलवद्दंष्ट्रिवाधनाः ॥ १८ ॥
एकोनानि तु पंचाशदाढकानि समादिशेत् ।

अश्वित्यां कुरुते यत्र वर्षणं नात्र संशयः ॥ १९ ॥
सस्यं शुभं भवत्यत्र पीड्यन्ते यवनाः शकाः ।
गान्धारिकास्तु काम्बोजाः पाञ्चालाश्च चतुष्पदाः ॥२०॥
एकोनसप्तति विन्द्यादाढ कानि न संशयः ।

Well-being, abundant food, and health, without the influence of
the five nights,
Strong-toothed creatures should be known, such as mice,
grasshoppers, and parrots. || 11 ||
Salty soil is known in the Varuna region and also causes distress
to crops,
Thieves are considered strong, and no one escapes their grasp. ||
12 ||
When God pours rain from the beginning of Purva Bhadrapada,
The crops then benefit from the rains, and it is Madhava's
blessing. || 13 ||
All grains are born, and the robbers become powerful,
When the leader is agitated, there is no influence of the ten nights.
|| 14 ||
Instruct to allocate ninety more units for the fields,
Cultivate the soil, and the prosperity of all shall increase. || 15 ||
Well-being, abundant food, and health without the influence of
the twenty nights,
One should know the divine beings as Bhadra Bahu has spoken. ||
16 ||

Direct sixty-four units for the Reva region,

All crops grow without a doubt. || 17 ||

The kings rise, and they oppose each other,

Those who are the main leaders shine with the power to crush the strong-toothed foes. || 18 ||

Allocate forty-nine units minus one,

In the Ashwini region, where there is no doubt about the rain. || 19 ||

The crops are auspicious here, the Yavanas and Shaka are oppressed,

Gandharikas, Kambojas, Panchalas, and quadrupeds are also affected. || 20 ||

Without a doubt, allocate sixty-nine units from the Vindhya region.

भरण्यां वासवचैव यदा कुर्यात्प्रवर्षणम् ॥ २१ ॥
सरिसृपास्तथा कीटा मरणं व्याधयस्तथा ।
सस्यं कनिष्ठं विज्ञेयं प्रजाः सत्राच दुःखिताः ॥ २२ ॥
आढकान्येकपंचाशत् कृतिकामु समादिशेत् ।
तदात्वपग्रहो ज्ञेयः सप्तविंशतिरात्रकः ॥ २३ ॥
द्वौ मासौ न तदा मेघाछिद्रं सस्यमुपद्रवः ।
निम्नेषु वापयेद्द्विज भयं चाग्नेर्विनिशिदेत् ॥ २४ ॥
आढकान्येकनवति रोहिण्यां वर्षणं यदा ।
तदा प्रवर्षणं विध्यात्सक दशाहकं ॥ २१ ॥
क्षेमं सुभिक्षमारोग्यं नैऋत्यां तु बहूदकं ।

स्थलेषु वापयेद्वीजं राज्ञोविजयमादिशत् ।। १६ ।।

मृगशीर्षे चाढकानां चतुःषष्टिं समादिशेत् ।
क्षेमं मुभिक्षयारोग्यं दंष्ट्रिणः प्रवलास्तथा ॥ २७ ॥

आढकानिनुद्वात्रिंशाद्रायां चापि निद्दिशेत् ।
बुभुक्षा मरणंव्याधि सस्यघातमुपद्रवः ॥ २८ ॥

श्रावणे प्रथने मासे वर्षित्वा न च वर्षति ।
प्रोपदं च वर्षित्वा शेषं कालं न वषति ॥ २९ ॥

आढकान्येव नवतिं विद्याञ्चैव पुनर्वसौ ।
सस्यं निष्पद्यते क्षिप्रं व्याधिश्च बलवान्भवेत् ॥ ३० ॥

द्विचत्वारिंशदपितुजानीयादाढका निवै ।
पुष्येण मंदवृष्टिं च निम्ने बीजानि वापयेत् ॥ ३१ ॥

When the rain comes from the direction of Vasava, it brings floods, ॥ 21 ॥

And the death of snakes, insects, and diseases,

The least growth of crops should be known, and the people are distressed. ॥ 22 ॥

One should predict a total of 151 rains,

Then the inauspicious period should be known to be 27 nights long. ॥ 23 ॥

For two months, there will be no rain, a gap in clouds, and crop disturbances,

In the lowlands, the Brahmin should perform water rituals, and the fear of fire should be dismissed. ॥ 24 ॥

When the rains come during the Rohini Nakshatra, amounting to 191,

Then one should predict a good rainfall for ten days. ‖ 21 ‖

Prosperity, ample food, and good health will be there, and the southwest will have abundant water,

In the lands, the Brahmin should perform water rituals, predicting the king's victory. ‖ 16 ‖

During Mrigashirsha Nakshatra, one should predict a total of 64 rains,

Prosperity, abundant food, good health, and powerful predators will be there. ‖ 27 ‖

One should predict a total of 32 rains during the Ray Nakshatra,

Hunger, death, diseases, and crop destruction will cause disturbances. ‖ 28 ‖

In the first month of Shravana, it rains but does not continue,

And after it rains for a while, it stops for the rest of the time. ‖ 29 ‖

One should know 90 rains in the Nakshatra of Vidya and again in Punarvasu,

Crops will quickly come forth, and diseases will become powerful. ‖ 30 ‖

One should predict 42 rains during the Pushya Nakshatra,

With mild rain, one should sow seeds in lowlands. ‖ 31 ‖

पक्कं चाश्वयुजिमासे पक्कं पौष्टपदे तथा ।
उपग्रहं विजानीयार बहुले प्रवर्षति ॥ ३२ ॥
चतुःषष्टिमादकानि तदा वर्षदि वासव ।
यदाश्लेषासु प्रथमं कुरुते च प्रवणम् ॥ ३३ ॥
सस्यघातं विजानीयात व्याधिभिशोदकेन च ।
साधवो दुःखिता ज्ञेयाः । प्रोपदेउपग्रहं ॥ ३४ ॥
मघायां मनवो द्रोणाः यदा तत्र प्रवर्षणम् ।
प्रजाश्च मुदिता ज्ञेया शुभंसस्यं च निर्दिशेत् ॥ ३५ ॥
फाल्गुनीषु च पूर्वामु यदा देवः प्रर्वपति ।
खारिं तदा दिशेद्वारि सुखं स्त्रीणां तदा भवेत् ॥ ३६ ॥
कुक्षिव्याधिश्च बलवान्वर्तते लोकनाशकृत् ।
अपग्रहश्च विज्ञेयः श्रावणे सप्तरात्रिकः ॥ ३७ ॥
उत्तरायां च फाल्गुन्दां पष्टि च नवतिं दिशेत् ।
आढकानि सुभिक्षं च क्षेममारोग्यमेव च ॥ ३८ ॥
दानशीलाच मनुजा धर्मशीलाच साधवः ।
अवग्रहं विजानीयात् कार्तिक द्वादशान्हिकम् ॥ ३९ ॥
।
पंचाशति विजानीयाद् हस्ते प्रवर्षणं यदा ।
तदानिम्नानि वाप्यानि जायते वर्षर्षचकं ॥ ४० ॥
संग्रामात्र प्रवर्धन्ते शिल्पिकानां मुखोत्तमाः ।
श्रावणे चाश्वयुजौ वा तथा कार्तिक एवं च ॥ ४१ ॥

In the month of Ashwin, when the moon is full, and likewise in Paush, know the presence of Upagrahas, causing heavy rainfall. (32)

When the god of rain, Indra, showers for sixty-four days, starting from the first day in Ashlesha, it indicates the onset of rainfall. (33)

Understand the destruction of crops due to diseases and flooding, and know that righteous people will suffer; this is the effect of the Upagrahas. (34)

When there is rainfall in Magha, people should be joyful, and it indicates good crop and prosperity. (35)

When the god of rain showers in the first half of Phalguni, there will be a scarcity of water; however, there will be happiness for women. (36)

A powerful epidemic causing abdominal illness will spread, destroying the world; know this as an effect of the Upagraha, occurring during the seven nights of Shravana. (37)

During the eleventh day of Uttarayana in Phalguna, predict ninety good days, with an abundance of food, peace, and good health. (38)

People who are generous and righteous should know the occurrence of the Avagraha during the twelfth night of Kartika. (39)

When there is rainfall in Hasta, know that fifty water reservoirs will be filled, and the cycle of rainfall will begin. (40)

The greatest artisans will flourish from the battles, during the months of Shravana, Ashwin, or Kartik. (41)

Understand the lunar phase month by month, ten days at a time, The thieves and the strong, they deceive and oppress the people. (42)

When twenty-two lunar days have passed, it is Chitra constellation and the rain begins,
Recognize the pattern then, this auspicious pattern, as Indra showers down. (43)

Plant seeds in low lands, avoiding barren places,
Understand the middle ground, as spoken by Bhadra-bahu. (44)

When thirty-two lunar days have passed, the pre-monsoon season begins,
Winds blow, rain falls, and the year proceeds without drought. (45)

In the Vishakha constellation, there's no doubt it is a harsh year,
All crops are affected, and trade is hampered. (46)

Understand the lunar phase, ten days into the month of
Praushthapada,
Undoubtedly, there is prosperity, abundance, and good health.
(47)

Understand the rain in Anuradha constellation to be harsh and
balanced,
Prosperity, abundance, good health, and the end of foreign threats.
(48)

Far away, travelers go, and virtuous people maintain their ways,
All crop yields are known, and fears are calmed. (49)

During Jyeshtha constellation, designate sixty-four lunar days,
Plant seeds in the ground, then abundant agriculture will thrive.
(50)

Know the harsh rain in the Mula constellation, which makes all
crops flourish,
Many roots are consumed, and thieves' activities increase. (51)

This has been told by Vyasa, and once more, wonder at the
arrangement.

द्वात्रिंशदाढकानि स्युर्नकपासे प्रवर्षणम् ॥ ५२ ॥
द्विगुणं च समक्षेत्रे च्चेषु त्रिगुणं स्मृतम्
शुभाशुभं ततो वाच्यं सम्यग् ज्ञात्वा यथाक्रम ॥ ५३ ॥
दशयुक्ताद्विकृतखतिथिरसाष्टदिग्विषयरामजल तिथिभिः ।
तिथिरसरसैश्च विरसाः सद्दशकृताः षविहीनाश्च ॥ ५४
जलषङ्कदशकसहिता जलरसयुक्ताः पड्ूनाश्च ।
विषयतिथिषङ्क सहिताश्चाश्विन्यादिषु जलद्रोणाः ॥ ५५ ॥
रविरवि सुतके तुपीडिते
क्षितितनयत्रिविधाद्भुताहते च ।
भवति च न शिवं न चापि वृष्टिः
शुभसहिते निरुपद्रवे शिवं च ॥ ५६ ॥
केचिद् यथाभिवृष्टं दशयोजनमण्डलं वदन्यन्ये ।
गर्गवासिष्ठपराशरमतेम तद् द्वादशान परम् ॥ ५७ ।
हस्तविशालं कुण्डकमधिकृत्याम्बुप्रमाणनिर्देशः ।
पञ्चाशत्पलमाढकमनेन मिनुयाज्जलंपतितम् ॥ ५८ ॥
येन धरित्री मुद्रा जनिता वा विन्दवस्तृणाग्रेषु ।
वृष्टेन तेन वाच्यं परिमाणं वारिणः प्रथमम् ॥ १९ ॥
येषु च भेष्वभिवृष्टं भूयस्तेष्वेव वर्पति प्रायः ।
यदिनाप्यादिषु वृष्टं सर्वेषु तदात्वनावृष्टिः ॥ ६० ॥

In these verses, the original author discusses various aspects of rainfall prediction and its impact on the environment.

32 pots of rainfall are considered auspicious in the nakshatra cycle. || 52 ||

In the case of even fields, it is doubled, while in odd fields, it is tripled.

Auspicious and inauspicious events should be declared after understanding them properly in order. || 53 ||

With the combination of ten, modified by the lunar day, the region's ruler, and the eight directions, the rainfall is determined by the lunar days.

The lunar days are matched with the corresponding tasteless ones, devoid of the number six. || 54 ||

Combined with the ten water-containing tastes and the water-containing lunar days, they are also the lunar days of destruction.

In the lunar mansions starting with Ashwini, there are water-filled pots accompanied by the taste-related lunar days. || 55 ||

When the sun is afflicted, and the earth's daughter is struck by the three types of afflictions,

Neither good nor rainfall occurs; but with auspiciousness and without any trouble, good things happen. || 56 ||

Some say that a circle of ten yojanas is where the rainfall occurs, while others have different opinions.

According to the views of Garga, Vasistha, and Parashara, the limit is twelve. || 57 ||

Taking a large-hand-sized pot, the amount of water is indicated.

With this, the water that has fallen is reduced by fifty-pala-pots. || 58 ||

By which the earth's watermark is created, or on the tips of the grass,

The amount of water from that rainfall should be declared first. || 19 ||

In which regions more rainfall occurs, there it usually increases.

If there is no rainfall in the beginning of the day, then there is no rainfall in all. || 60 ||

आषाढादिषु वृष्टेषु योजनद्वादशात्मके ।
प्रवृष्टे शोभनं वर्षं वर्षाकाले विनिर्दिशेत् ।। ६१ ।।

In the months starting from Ashadha, during the twelve yojana rains,

One should predict a prosperous year when there is abundant rainfall. || 61 ||

रोहिणी योगाध्यायः ५

प्राजेशमापाढतमिस्रपक्षे क्षमाकरेणोपगतं समीक्ष्य ।
वक्तव्यमिष्टं जगतोऽशुभं वा, शास्त्रोपदेशाद्ध हचिन्तकेन ॥ १॥

नगरादुपनिष्क्रम्य दिशां प्रागुत्तरां शुचिः ।
विविक्ते प्रस्थले देशे देवतायतनेऽपि वा ॥ २ ॥
राज्ञा नियुक्तो दैवज्ञः कृतशौचोजितेन्द्रियः ।
निमित्त कुशलो धीरः शुक्लांवरसमावृतः ॥ ३ ॥
ततोऽष्टम्यां परे यस्मिन् दिने संयुज्यते शशी ।
प्राजापत्येन च ततो निमित्तान्युपलक्षयेत् ॥ ४ ॥
लक्ष्णां पताकामसितां विदध्याद्
दण्डप्रमाणां त्रिगुणोच्छ्रितां च ।
आदौ कृते दिग्ग्रहे नभस्वान्
ग्रास्तया योगगते शशांके ॥ ५ ॥
तदहश्चोदयादूर्ध्व चतुर्षा हो विभज्य च ।
हिताहितार्थं मासानां चतुर्णामुपलक्षयेत् ॥ ६ ॥
दिनार्द्ध शुभदो वायुमासौ तत्र वर्षति ।
'चतुर्भागेन मासं तु शक्रोऽत्यर्थं प्रवर्षति ॥ ७ ॥
पूर्वे चै वार्द्धदिवसे पूर्वौ मासौ तु वर्षति ।

Rohini Yoga Chapter 5

Observe the conjunction of Prajapati (Jupiter) with the waning
moon in the Kshama constellation,
And based on the teachings of scriptures, discern whether it is
auspicious or inauspicious for the world. || 1 ||

Leaving the city, head towards the northeast direction, pure,

In a secluded, open area or even in a temple dedicated to deities. || 2 ||

Appointed by the king, the astrologer should be pure, self-controlled, and have his senses in check,
He should be skilled in omens, courageous, and attired in white garments. || 3 ||

Then, on the day after the eighth lunar day when the moon is in conjunction,
He should observe the omens associated with Prajapati (Jupiter). || 4 ||

Fix a black flag of standard size with three times the height of a staff,
At the beginning of the eclipse, when the moon is in the sky and being devoured by the shadow. || 5 ||

Then, divide the day into four parts and observe them,
To determine the positive and negative effects for the four months. || 6 ||

During the first half of the day, if there is a wind blowing, it will rain in those months,

And for a quarter of the month, Indra will bring an excessive downpour. || 7 ||

In the earlier part of the increasing day, it will rain in the first two months

अहस्तु पश्चिमे भागे पश्चिमौ द्वौ तु वर्षति ॥ ८ ॥
अथ पूर्व व्यतिक्रम्य भागं तत्पश्चिमं ततः ।
मध्यान्हे वाति चेद्रायुर्मध्यौ मासौ तु वर्षति ॥ ९ ॥
भाद्रपदोऽश्वयुक् चैव मासावेतौ तु मध्यमौ ।
एनयोरपि निर्देश्या वर्षारात्रस्य संपदः ॥ १० ॥
वायन्तं मारुतं चापि यो वायुः प्रतिवायति ।
तत्र यो बलवान् वायुस्तस्यैव फलमादिशेत् ॥ ११ ॥
योगे धनुद्धता वाता हूलादयन्तः सुखप्रदाः ।
प्रदक्षिणाः श्रेष्टतमाः पूर्वपूर्वोत्तरा इति ॥ १२ ॥
शान्तपक्षिमृगरावितादिशो, निर्मलं वियदनिन्दितोऽनिलः
शस्यते शशिनि रोहिणीगते, मेघमारुतफलानि वच्स्यतः ॥ १
कचिदसितसितैः सितैः कचिच्च
क्वचिदसितैर्भुजगैरिवांब्रुवाहैः ।
वलितजठरपृष्ठमात्र दृश्यैः
स्फुरिततडिद्रसनैर्वृतं विशालैः ॥ १४ ॥
विकसित कमलोदरावदातै-
ररुणकरद्युतिरंजितोपकंठैः ।
छुरितमिव वियद् घनैर्विचित्रे-

मधुकर कुंकुमकिंशुकावदातैः ॥ १५ ॥
असितघननिरुद्धमेव वा
चलिततडित्सुरचापचित्रितम् ।

द्विपमहिपकुलाकुलीकृतं

वनमिव दाबपरीतमम्बरम् ॥ १६ ॥
अथवाञ्जनशैलशिलानिचयप्रतिरूपधरैः स्थगितं गगनम् ।
हिममौक्तिक शंख शशांककरद्युतिहारिभिरम्बुधरैरथवा १७
तडिद्वैमकक्ष्यैर्वेलाकाग्रदन्तैः
स्रवद्वारिदानैश्वलत्प्रान्तहस्तैः ।
विचित्रेन्द्रचापध्वजे । च्छाय शौभ
स्तमालालिनीलैर्वृतं चान्दनागैः ॥ १८ ॥
आषाढवहुलपक्षे शिशिरकरे रोहिणीसमायुक्ते ।
यदि गगनममममसन्ततीक्ष्णरश्मिः सहस्रांशुः ॥ १९ ॥
सलिल गुरुनम्र जलधरत डिल्लतालोलरंज्जितदिगतः ।
अमितमलभेक चातककादम्बविमिश्रमाकाशम् ॥ २० ॥
क्षितितनयरविजरहितः स्फटिकनिभश्चंद्रमा निरुत्पातः ।
मरुतश्च पूर्वपूर्वोत्तरोत्तराः शान्तमृगविहगाः ॥ २१ ॥
निगदितरूपैर्जलधरजालै-
रूयहमवरुद्धं द्वहमथवाहः
यदि वियदेवं भवति सुभिक्षं
मुदितजना च प्रचुरजला भूः ॥ २२ ॥
छिन्नमूलाश्च वृक्षाश्च शुष्का बाष्पाकुलीकृताः ।
पापसत्वानुकाराश्च मेघाः पापफलप्रदाः ॥ २३ ॥

In the west, after the sun descends, for two months it rains. || 8 ||

Crossing the eastern part, then to the western part thereafter,

If the wind blows at noon, for two middle months it rains. || 9 ||

Bhadrapada and Ashvayuk are the middle months,

For both of these, the wealth of the rainy season is prescribed. || 10 ||

The wind that counters the moving and stationary winds,

The strongest of those winds, should be designated as the fruit. || 11 ||

In the yoga, the winds are pleasing during the bowstring drawn time,

Circumambulating clockwise, the most excellent are from the east and northeast. || 12 ||

In the tranquil bird-animal constellation, a cloudless sky, with a wind without blemish,

When the Moon is in Rohini, I will tell of the fruits of clouds and winds. || 1 ||

Sometimes white, sometimes dark, like serpent brows,

With patterns like the back of a striped belly, revealing vast lightning. || 14 ||

With expanded lotus-like hearts, reddish hands glowing around the neck,

The sky is cut, as if by a colorful array of bees, Kunkuma, and Kinshuka flowers. || 15 ||

Or it seems like a dense cloud, obstructed by darkness,

The sky adorned with the pictures of the moon's bow and lightning,

Filled with the commotion of elephants, and surrounded by forests. || 16 ||

Or the sky appears motionless, with mountain-like cloud shapes,

Or adorned with strings of pearls, conch shells, and moonbeams. || 17 ||

With streaks of lightning, like the teeth of a crocodile,

With the edges of the downpour, and the flags of Indra's bow,

Surrounded by dark green sandalwood trees. || 18 ||

During the waxing phase of Ashadha, with the cold Moon in Rohini,

If the sky is filled with sharp rays of the thousand-rayed Sun. || 19 ||

The water-laden, humble clouds, with their playful movements,

Fill the sky with the mixture of sounds from the Cuckoo and Kadamba birds. || 20 ||

The Earth's child, devoid of the Sun's rays, and the clear moon without disturbances,

The winds are calm, from the east and northeast, with quiet animals and birds. || 21 ||

With the described forms of the net of clouds,

If the sky is thus obstructed, the Earth is abundant with water and people are happy. || 22 ||

Uprooted trees and dry, tear-filled clouds,
Imitating sinful beings, the clouds yield sinful fruits. || 23 ||

विगतघने वा वियति विवस्वा-
नमृदुमयूखः सलिलकदेवम् ।
सर इव फुल्लं निशि कुमुदाढ्यं
खमुडविशुद्धं यदि च सुवृष्टयै ॥ २४ ॥
पूर्वोदभूतैः सस्यदिष्पतिरन्दै-
राग्नेयाशा संभवैरग्निकोपः ।
याम्ये सस्यं क्षीयते नैर्ऋतेऽई
पश्चाज्जातैः शोभना वृष्टिरब्दैः ॥ २९ ॥
वायव्योत्थैवतिवृष्टिः कचिच्च
पुष्टा वृष्टिः सौम्यकाष्ठा समुत्थैः ।
श्रेष्ठं सस्यं स्थाणुदिक संप्रवृद्धै-
वायुचैवं दिक्षु धत्ते फलानि ॥ २६ ॥
उल्कानिपातास्तडितोऽशनिश्र
दिग्दाह निर्घातमही प्रकंपाः ।
नादा मृगाणां सपतत्रिणां च
ग्राह्या यथैवाम्बुधरास्तथैव ॥ २७ ॥
ग्रहान् सनक्षत्रगणान् समालिखेत्
सधूपपुष्पैबलिभिश्च पूजयेत् ।
सरत्नतोयौषधिभिश्चतुर्दिशं
तरुप्रवालापिहितैः सुपूजितैः ॥ २८ ॥

अकालमूलैः कलशैरलंकृतं
कुशास्तृतं स्थंडिलमावसेद्विजः ।
आलभ्य मंत्रेण महात्रतेन
बीजानि सर्वाणि निधाय कुंभे ॥ २९ ॥
लाव्यानि चामीकरदतोये-
होमो मरुद्वारुण सोममन्त्रैः ।
नामांकितैस्तैरुदयादिकं भैः
प्रदक्षिणं श्रावणमासपूर्वैः ॥ ३० ॥
पूर्णैः समासः सलिलस्य दावा
सुतैरवृष्टिः परिकल्प्यमूनैः ।
अन्यैश्व कुंभैर्नृपनामचिन्है-
देशांकितैश्चाप्यपरैस्तथैव ॥ ३१ ॥
उदगपि च तुहिनकिरणः पूर्वोत्तरतोऽथवा स्थितः प्राच्याम् ।
यदि भवति तदा वसुधा भवति विवृद्धा प्रहृष्टजना ॥ ३२ ॥
उपसर्गोऽनलादिस्थे याम्याशासंस्थिते शकटके च ।
किं कष्टैस्तैरुक्तैः श्रुतमात्रैयैः कृशो भवति ॥ २३ ॥
क्रिमिशुकशलभादिभयं नैऋत्यां नातिपुष्टिरपरेण ।
वायव्याशा संस्थे मध्यं संस्ये कुमुदनाथे ॥ ३४ ॥
सतारागणमध्ये वै या तारा दीप्तिमत्तरा ।
योगतारेति सा प्रोक्ता नक्षत्राणां पुरातनैः ॥ ३५ ॥

विक्षेपशद्वितीयादधिको चाभ्रभुजभागे
यस्य ग्रहस्य याम्यो भिनति शकटं स रोहिण्याः ॥ ३६ ॥
ताडयेत् यदि च योगतारका-

मावृणोति वपुषा यदापि वा ।
ताडने भय मुशन्ति दारुणं.
छादने नृपवर्षोऽगनाकृतः ॥ ३७ ॥
प्राक् प्रवेशे तु यूथस्य पुरतो वृषभो यदा ।
प्रवेशो कृष्णवर्णो वा पशुहुजलप्रदः ॥ ३८ ॥
कृष्णा तु गौः सुभिक्षाय क्षेमारोग्याय चोच्यते ।
गौर्यामथ च नीलायां मध्याः सस्यस्य संपदः ॥ ३९ ॥
अनावृष्टिकरी श्वेता वाताय कपिला स्मृता ।
पाटला संस्वनाशाय रोगाय करटा स्मृता ॥ ४० ॥
एकदेशाय शबला चित्रं चित्रा तु वर्षति ।
पाण्डुरा मध्यमांगी वा ग्रीष्मधान्यविवर्द्धिनी ॥ ४१ ॥
कपिला पश्चिमं वर्षं शोणा वग्रे प्रवर्षति ।
लक्षयेतु निमित्तानि ग्रामद्वारे विचक्षणः ॥ ४२ ॥
दृश्यते न यदि रोहिणी युत-
चंद्रमा नभसि तोयदावृते ।
रुग्भयं महदुपस्थितं तदा
भूव भूरिजलसस्यसंयुता ॥ ४३ ॥

स्वातावणादास्वथ रोहिणीपु, पापग्रहा योगगता न शस्ताः
आह्यं तु योगद्वयमप्युपोष्य, यदाधिमासो द्विगुणी करोति ॥ ४४ ॥

In the absence of dense clouds, or when the sky is clear,
The soft rays of the sun illuminate the water-bearing god.
Like a lake full of blooming lotuses at night,
The sky becomes purified, if it is well-rained upon. || 24 ||

40

With the emergence of crops, there's the prosperity of grains,
And with the Agneya (southeastern) direction, the fire's fury
arises.
In the Yamyā (southwestern) direction, the crops decrease; in the
Nairṛtī (northwestern), there's growth,
And with the rainfall in the latter half, there's splendid growth of
crops. || 29 ||

With the wind blowing, there's abundant rainfall,
And with the nourishing rain, the Soma plants grow.
In the Sthāṇu (northeastern) direction, the best crops flourish,
As the wind also bears fruits in all directions. || 26 ||

Meteors, streaks of lightning, and thunderbolts,
Burning of directions, the shaking of the earth.
The sounds of animals and the rustling of leaves,
Are all to be grasped just like water-bearing clouds. || 27 ||

One should list the planets and constellations,
And worship them with offerings of incense and flowers.
With gemstones, water, and herbs, in all four directions,
Worship them with trees, shoots, and well-covered leaves. || 28 ||

Adorned with sacred grass and water pots,
The Brahmin should sit on the Kusha-grass-covered platform.
By chanting the great mantra and invoking it,
Place all the seeds in the pot. || 29 ||

With the water of the Amīkara river,
Perform the Homa with Marut, Varuna, and Soma mantras.
With the named and mentioned directions, from the east onwards,
Perform the Pradakshina in the month of Shravana. || 30 ||

With the fullness of the water, the fire is extinguished,
With the verses, the rain is calculated.
With other pots marked with the names of kings,
And with the symbols of the regions, do the same. || 31 ||

If the frosty rays of the sun are in the east-northeast,
Then the earth flourishes, and the people rejoice. || 32 ||
When the eclipse occurs in the fire and other elements, in the southwest,
With the mentioned difficulties, the earth becomes weak. || 23 ||

There's fear of worms, parrots, locusts, and the like in the south,
And not much growth in the other directions.
In the northwestern direction, there's moderate growth of crops,

In the region of the lord of lotuses (the moon). || 34 ||

Among the countless stars, the one that shines the brightest,
Is called the Yoga-tārā, as said by the ancients among the
constellations. || 35 ||

36. When the second of the six perturbations is greater and the
southern planet breaks the cart of Rohini constellation, it is
significant.

37. If it strikes the pole star or covers it with its body, even
momentarily, fear of a terrible calamity is dispelled when it
strikes, and the king's reign is untroubled during the covering.

38. When a bull is in front of the group before entering, or the
entrance features a black color, or when a cow with a pot of water
is offered, it's considered auspicious.

39. A black cow is recommended for good harvests, welfare, and
health; for white and blue cows, there is an abundance of crops.

40. A white cow brings drought, a brown cow is associated with wind, a red cow is for noise reduction, and a black cow is considered to bring disease.

41. A spotted cow brings rain to one region, while a cow with an extraordinary appearance causes rain. A pale cow or one with a middle stripe increases the summer harvest.

42. A brown cow brings rain in the western region, and a red cow causes rain in the eastern region. A wise person must observe these signs at the village gate.

43. If Rohini, combined with the moon, cannot be seen in the sky due to clouds, then a great danger arises, accompanied by a massive flood and loss of crops.

44. From the new moon to the full moon in Rohini, the sinful planets in conjunction are not harmful. However, one should fast during the two conjunctions when the additional month doubles the effect.

स्वातियोगाध्यायः ६

यद्रोहिणीयोगफलं तदैव
स्वातावपाढासहिते च चंद्रे ।
आषाढशुक्ले निखिलं विचिन्त्यं
योऽस्मिन् विशेषस्तमहं प्रवक्ष्ये ॥ १ ॥
स्वातियोगे यदा युक्ते पूर्वरात्रे प्रवर्षति ।
ग्रीष्मशारदसंपन्नां तां समामभिनिर्दिशेत् ॥ २ ॥
रात्रेद्विभागमाश्रित्य स्वातियोगेऽभिवर्षति ।
सम्पदो मुद्रमाषाणां तियानां चावधारयेत् ॥। ३. ॥
त्रिभागशेषे शर्वर्याः स्वातियोगेऽभिवर्षति ।
ग्रैषम्यं संपद्यते सस्यं शारदं तु विनश्यति ॥ ४ ॥
अहस्तु प्रथमे भागे वर्ष क्षेमदृष्टये ।
द्वितीये शोभना वृष्टिर्वहुसस्य सरीसृपाः ॥ ५ ॥
अहनस्तृतीये भागेतु मध्यमां कुरुते समाम् ।
अहोरात्रं यदावृष्टः स्वातियोगे पुरंदरः ॥ ६ ॥
तदा तु चतुरो मासान् सर्वान् वर्षति वासवः ।
एवं दिनविभागेन दृष्टिः प्रोक्ता पुरातनैः ॥ ७ ॥
सममुत्तरेण तारा चित्रायाः कीर्त्यते नृपवत्सः ।

Swati Yoga Chapter 6

1. The results of Rohini Yoga are similar to those of Swati Yoga with the moon in conjunction with Apah (Purva Phalguni). In the Shukla Paksha of Ashadha, I will describe the unique effects to be considered in this context.

2. When Swati Yoga occurs in the first half of the night and it rains, it indicates a prosperous summer and autumn season.

3. If it rains during Swati Yoga in the first half of the night, one should expect prosperity in the form of coins and grains.

4. If it rains during Swati Yoga in the remaining third of the night, a good crop is expected in the summer season, but the autumn crop may be destroyed.

5. If it rains during the first part of the day, it is considered auspicious for prosperity. In the second part, a beautiful rain brings an abundance of crops and serpents.

6. During the third part of the day, the rain results in a moderate outcome. When it rains during both day and night in Swati Yoga, Purandara (Indra, the god of rain) is present.

7. Then, for all four months, Vasava (Indra) brings rain. In this way, the ancient vision is described based on the divisions of the day.

8. Swati, the star in the northern direction, is hailed as the king's star along with Chitra.

ऊकुराश्व वल्यस्सद्यो वर्षाय कीर्त्यते ।
स्निग्धाः समसितरेखा यथाभ्रवृन्दानि कल्पितान्येव ॥ ८ ॥
उदयशिखरिसंस्थो दुर्निरीक्ष्योऽतिदीप्त्या ।
तकनकनिकाशः स्निग्धवैदूर्यकान्तिः ॥
तदहनि कुरुतेऽभः तोयकाले विवस्वान्
प्रतपति यदि चोच्चैः खंगतोतीव तीक्ष्णः ॥ ९ ॥
नेच्छन्ति विनिर्गमं गृहाद्
धुन्वन्ति श्रवणान् खुरानपि ।
पशवः पशुवच्च कुक्कुरा
यद्यभः पततीति निर्दिशेत् ॥ १० ॥
स्तनितं निशि विद्युतो दिवा
रुधिरनिभा यदि दण्डवत्स्थिताः ।
पवनः पुरतश्च शीतलो
यदि सलिलस्य तदागमो वदेत् ॥ ११ ॥
यदास्थिता गृहपटलेषु कुक्कुरा
रुदन्ति वा यदि विततं वियन्मुखाः ।
दिवा तडिद्यदि च पिनाकदिग्भवा
तदा क्षमा भवति समैव वारिणा ॥ १२ ॥
यदि तितिरिपत्रनिभं गगनं
मुदिताः प्रवदन्ति च पक्षिगणाः ।
उदयास्तमये सवितुर्द्युनिशं

विसृजति घना न चिरण जलम् ॥ १३ ॥

प्रावृषि शीतकरो भृगुपुत्रात्
सप्तमराशिगतः शुभदृष्टः ।
सूर्यसुतान्नवपंचम वा
सप्तमगश्च जलागमनाय ॥ १४ ॥

यद्यमोषकिरणाः सहसगो-
रस्तभूधरकरा इवोच्छ्रिताः ।
भूसमं च रसते यदाम्बुद-
स्तन्महद्भवति वृष्टिलक्षणम् ॥ १५ ॥

मयूरशुक चाषचातक समानवर्णा यदा
जपाकुसुमपकंजद्युतिमुखाश्च सन्ध्याघनाः ।
जलोर्मिनगनक्रकच्छपवराहमीनौपमाः
प्रभूतपुट सञ्चया न तु चिरेण यच्छन्त्यपः ॥१६॥

शक्रचापपरिघप्रतिसूर्या रोहितोऽथ ताडितः परिवेषः ।
उद्गमास्तसमये यदि भानोरादिशेत्प्रचुरमंबु तदाशु ॥ १७॥

पर्यन्तेषु सुधाशशांकधवला मध्येऽञ्जनालित्विषः
स्निग्धा नैकपुटाः क्षरज्जलकणाः सोपानविच्छेदिनः ।
माहेन्द्रीप्रभवाः प्रयान्त्यपरतः प्राग्वांबपाशोद्भवा
ये ते वारिमुचस्त्यजति न चिरादंभः प्रभूतं भुवि ॥ १८ ॥

बलवत्सु महद्वर्षमल्पेष्वल्पवृशीकरम् ।
मध्येषु मध्यमं ब्रूयान्निमितेषु निमित्तवित् ॥ १९ ॥

In these verses, the poet describes various indications of impending rain and the beauty of nature.

8. The Ukurashva and Valyassa frogs are singing, heralding the arrival of rain. The streaks of clouds appear smooth and well-formed, as if carefully crafted.

9. At sunrise, the rays are hard to discern due to their intense brightness, like the brilliance of polished gems. If the sun's rays become sharper as the day progresses, it indicates the arrival of rain.

10. Animals, like cows and dogs, refuse to leave their shelters and even rub their ears against the ground. They sense the impending rain.

11. If there's thunder at night and lightning during the day, and if the wind blows cool air ahead of the rain, it indicates the imminent arrival of rain.

12. When dogs sit on the floor of their homes and cry, or birds spread their wings wide, and if lightning appears during the day, then forgiveness is granted as the rain comes.

13. When the sky takes on the hue of a tittiri bird's wing, and birds sing joyously, the sun sets and rises, and clouds release water without delay.

14. During the rainy season, a favorable view from the seventh constellation, the son of Bhrigu, brings rain. The sun's ninth and fifth sons also indicate the arrival of rain.

15. When the rays of the sun appear like the horns of a cow, and when the clouds touch the ground, it is a sign of heavy rainfall.

16. When clouds resemble peacocks, parrots, and cuckoos, and when they have the colors of hibiscus flowers and lotus petals, they bring abundant water without much delay.

17. If, at the time of sunrise or sunset, the sky displays the colors of Indra's bow, a mace, or a discus, it predicts the arrival of heavy rain.

18. Clouds originating from the great mountains and the eastern region bring continuous rainfall. When they release water, it soon covers the earth in ample amounts.

19. The one who knows the signs should predict a heavy downpour in strong indications, light rain in weak indications, and moderate rain in average indications.

उल्कानिर्घातभूकंपपांशुवर्षाणि केतवः
अपसव्या ग्रहाथैव नित्यं वर्षासु वर्षदाः ॥ २० ॥
प्रायोग्रहाणामुदयास्त काले
समागमे मण्डलसंक्रमे च ।
पेक्षक्षये तीक्ष्णकरायनान्ते
वृष्टिर्गतर्फे नियमेन चार्द्रा ॥ २१ ॥
अग्रतः प्रष्टतो वापि ग्रहाः सूर्यावलंविनः ।
यदा तदा प्रकुर्वन्ति महामेकार्णवाभव ॥ २२ ॥
समागमे पतति जलं ज्ञशुक्रयो-
ईजीवयोर्गुरुसितयोश्च संगमे ।
यमारयोः पवन हुताशजं भयं
ह्यदृष्टयोरसहितयोथ हैः ॥ २३ ॥

Comets, lightning, earthquakes, and showers of dust are the signs,
Inauspicious planets, indeed, always bring rain during the rainy
season. || 20 ||
At the time of the rising of the planets, during their conjunction,
At the crossing of the orbits and at the end of the sharp and
elongated phase,
Rain falls on the waning day, as per the rule, it is wet. || 21 ||
Whether in front or behind, the planets follow the Sun,
Whenever they do, they cause the great ocean to rise. || 22 ||

During the conjunction, water falls, when Jupiter and Venus,
Mercury and the Sun, Saturn and Mars meet,
The wind and fire bring fear, when Rahu and Ketu are together. ||
23 ||

प्रकीर्णकशुभाशुभयोगाध्यायः ८

अहोरात्रस्य यः सन्धिः सा च सन्ध्या प्रकीर्तिता ।
द्विनाडिका भवेत्साधुर्यावदा ज्योतिदर्शनम् ॥ १ ॥
मृगशकुनिपवनपरिवेषपरिधपरिघाभ्रवृक्षमुरचापैः ।
गन्धर्वनगर रविकरदण्डरजः स्नेहवर्णैश्च ॥ २ ॥
अपसव्ये संग्रामः सव्ये सेनागमः शान्ते ।
मृगचक्रे पवने वा सन्ध्यायां मिश्र वृष्टिः ॥ ३ ॥

Chapter on the scattered auspicious and inauspicious
combinations: 8
The junction of day and night is called the twilight,
Two and a quarter nadis are considered good, as long as the stars
are visible. || 1 ||
With deer, birds, wind, clouds, lightning, trees, and rainbows,
The celestial city, the rays of the Sun, the colors of affection. || 2 ||
On the left side, there is a battle; on the right side, the army is at
peace,
In the deer circle or the wind, the twilight brings mixed rain. || 3 ||

गृहतरुतोरणमथने सपांशुलोष्टोत्करेऽनिले प्रबले ।
भैरवरावे रूक्षे खगपातिनि च शुभ सन्ध्या ॥ ४ ॥
सन्ध्या काले स्निग्धा दण्डतडिन्मत्स्यपरिधिपरिवेषाः ।
सुरपतिचापैरावृतरविकिरणाश्चाशु वृष्टिकराः ॥ ५ ॥
उद्यातिनः प्रसन्ना ऋजवो दीर्घाः प्रदक्षिणावर्ताः ।
किरणाः शिवाय जगतो वितमस्के नभसि मानुमतः ॥ ६ ॥
शुक्लाः करा दिनको दिव्यादिमध्यान्तगामिनः स्निग्धाः ।
अव्युच्छिन्ना ऋजवो वृष्टिकरास्ते त्वमोघाख्याः ॥ ७ ॥
कल्माषवभ्रुकपिला विचित्रमाजिष्टहरितशवलाभाः ।
त्रिदिवानुबंधनोये स्वष्टयेऽल्पभदास्तु ॥ ८ ॥
बन्धुजीवनिकाशेन तपनीयनिभेन वा ।
उदये रजसा सूर्यः संवृतः शखमावहेत् ॥ ९ ॥
शंखचूर्णनिकाशेन रजसा संवृतो रविः ।
राज्ञो विजयमाख्याति वृद्धिं जनपदस्य च ॥ १० ॥
रविकिरणजलदमरुतां संघातो दंडवत्स्थितो दंडः ।
स विदिस्थितो नृपाणां शुभो दिक्षु द्विजादीनाम् ॥ ११ ॥
दधिसदृशाभ्रो नीलो भानुच्छादी खमध्यगोऽभ्रतरुः ।
पीतच्छुरिताथ घना घनमूला सुघनवृष्टिकराः ॥ १२॥
कुवलयवैद्यवुजकिञ्जल्कामा प्रभञ्जनोन्मुक्ता ।
सन्ध्या करोति दृष्टि रविकिरणोद्भासिता सद्यः ॥ १३ ॥

In the powerful wind, crushing doorways and trees, with dust and clumps of earth flying,

In the fierce roar of Bhairava and the harsh cry of the lord of birds, auspicious twilight emerges. || 4 ||

At the time of twilight, smooth and adorned with fish-filled ponds,
The rays of the lord of gods, swiftly covered by clouds, bring rain. || 5 ||

The rising, pleased, straight and long rays moving clockwise,
Shining on the world, reach out to the sky in obeisance to Shiva. || 6 ||

The bright, divine, smooth and uninterrupted rays, reaching the beginning and the end,
O unobstructed ones, you bring rainfall, being welcomed by all. || 7 ||

With dark, curved brows and a variety of reddish-brown hues,
May those who follow the heavens bring well-being and little harm. || 8 ||

By the reddish hue of the bandhujiva flower or the heat of the sun,
At sunrise, may the sun be covered with dust and carry forth its rays. || 9 ||

Covered by the dust of the Shankha-churna, the sun,
Proclaims the victory of the king and the prosperity of the
kingdom. || 10 ||

The confluence of the sun's rays, clouds, and wind stands like a
staff,
Auspiciously placed in the directions for kings and the learned
alike. || 11 ||

Resembling curd, reddish and blue, the sun is hidden by a cloud
in the sky,
With a yellowish hue, the clouds are dense and have a copious
source of rain. || 12 ||

Released by the wind, desiring the petals of the water lily, the
trumpet flower, and the Kinjalka,
Twilight is revealed, illuminated by the sun's rays at the moment
of sight. || 13 ||

अशुभाकृतिघनगन्धर्वनगरनीहारधूमपांशुयुता
।
प्रावृषि करोत्यवग्रहमन्यतौं शस्त्रकोपकरी ॥ १४ ॥

वसन्ते मधुवर्णाभाऽथवा रुधिरसन्निभा ।
ग्रीष्मे श्वेता रजोध्वस्ता पांशुवर्णा च शस्यते ॥ १९ ॥
नीललोहित शुक्लामा सन्ध्या वर्षासु वार्षिका ।
मञ्जिष्टवर्णा शरदि पीयूषाभा च शस्यते ।। १६ ।।
हेमन्ते बभ्रुवर्णाच पिंगला चापि पूजिता ।
शिशिरे शोणवर्णाभा सन्ध्या क्षेमसुखप्रदा ॥ १७ ॥
स्निग्धाला सप्रभा नाकुलापि वा
सन्ध्या यथर्तुवर्णामा शान्तद्विजमृगा शुभा ॥ १८ ॥
सितसितान्त धनावरणं रखे-
भवति वृष्टिकरं यदि सव्यतः
यदि च वीरण गुल्मनिभैर्धनै-
दिवसभर्तुरदीप्त दिशुद्भवैः ।। १९ ।।
उभयपार्श्वगतौ परिधी खेः
प्रचुरतोयकरौ वपुषान्वितौ ।
अथ समस्त ककुप्परिचारिणः
परिधयोस्ति कणोऽपि न वारिणः ॥ २० ॥
प्रति सूर्यः शक्रधनुर्दण्डकः परिवेषणम् ।
।
तथैरावतमत्स्यास्त्र स्निग्धा ये चाक रश्मयः ॥ २१ ॥

In the presence of inauspicious shapes, thick with the scent of
Gandharva cities, adorned with mist, smoke, and dust,
In the rainy season, she wields a weapon and becomes angry,
showing no other restraint. || 14 ||

In spring, she is of a honey-like hue, or perhaps of a blood-like color,
In summer, she is white and free from dust, and takes on a sandy hue when the crops ripen. || 19 ||

In the rainy season, Sandhya appears as blue, red, and white,
In autumn, she is of the Manjistha color and has a milky appearance. || 16 ||

In the winter, she is of a tawny color, and also revered as being yellowish,
In the dewy season, she is of a reddish hue, and Sandhya bestows comfort and happiness. || 17 ||

With a smooth, radiant, and charming demeanor,
Sandhya is of the appropriate color according to the season,
pacifying the twice-born and the animals, and is auspicious. || 18 ||

When the silver-tailed cloud on the left side provides a covering,
It brings rain, but if it is to the right, the day is filled with intensely shining directions. || 19 ||

On both sides of the sky, there are ample water-bearing clouds present,

Yet, in the midst of all these clouds, there is not a single drop of water. || 20 ||

The sun moves in opposition to Indra's bow and its quiver,
And so do the smooth rays and the fish-shaped constellations. || 21 ||

विद्युतो भूरिकाराश्च वर्णा ये च प्रदक्षिणाः ।
संध्यासु यदि दृश्यन्ते सद्यो वर्षणलक्षणम् ॥ २२ ॥
प्राचातत्क्षणमेव नक्कमपरा सन्ध्या व्यहाद्वा फलम् ।
सप्ताहात् परिवेषरेणुपरिघाः कुर्वन्ति सद्यो न चेन् ॥
तद्वत्सूर्य करेन्द्रकार्मुकताडित्प्रयकमघानिला-
स्तस्मिन्नेवादिनेष्टमेऽथ विहगाः सप्ताहपाका मृगाः ॥ २३ ॥
संमूर्छिता रवीन्द्रोः किरणाः पवनेन मण्डलीभूताः
नानावर्णाकृतयस्तन्वभ्रे व्यान्नि परिवेषाः ॥ २४ ॥
सितपीतेन्द्रनीलाभा रक्तकापोत वम्रवः
शबला बर्हिवर्णाश्च विज्ञेयास्ते शुभप्रदाः ॥ २५ ॥
ऐन्द्रयाम्याप्यनैर्ऋत्यवारुणाः सौम्यवद्विजाः
दृश्यादृश्येन भावेन वायव्यः सोऽपिकष्टदः ॥ २६ ॥
धनदः करोति मेचक मन्योन्यगुणाश्रयेणचाप्यन्ये ।
प्रविलीयते मुहुर्मुहरल्प फलः सोऽपि वायुकृतः ॥ २७ ॥
शिशिरे चाषवर्णश्च वसन्ते शिखिसंनिभः ।
ग्रीष्मे रजतसंकाशः प्रावद् तैलसमप्रभः ॥ २८ ॥
गोक्षीर सदृशः शस्तः परिवेषः शरत्स्मृतः
हेमन्ते जलसंकाशः स्वकाले शुभदः स्मृतः ॥ २९ ॥

वर्णेनैकेन यदा बहुल: स्निग्धः क्षुराभ्रकाकीर्णः
स्वच सद्यो वर्षे करोति पीतश्च दीप्तार्कः ॥ ३० ॥
दिवामूर्ये परिवेषो रात्रौ चन्द्रे यदा भवेत् ।

When there are numerous flashes of lightning, and colors appear in circles,

If seen at twilight, it is a sign of imminent rain. || 22 ||

At the very moment of twilight, if the western sky is clear or if a halo is seen,

In a week, dust particles carried by wind will cause the halo, if not immediately. || 23 ||

The sun's rays, scattered by the wind, form circular patterns,

Creating various-colored streaks across the sky. || 24 ||

White, yellow, indigo, reddish, dove-colored, and reddish-brown,

These colors are auspicious and should be recognized. || 25 ||

The eastern, western, northern, and southern winds, and the gentle breezes,

Can be seen or unseen, and the wind from the northwest is also difficult. || 26 ||

The wind brings wealth, and others rely on each other's virtues,

The small result disappears again and again, and it is also caused by the wind. || 27 ||

In winter, the colors are pale; in spring, they resemble peacock feathers.

In summer, they shine like silver and have an oil-like glow. || 28 ||

In the rainy season, the halo is like cow's milk; in autumn, it is remembered.

In the cold season, it resembles water, and at its own time, it brings good fortune. || 29 ||

When a single color is abundant, smooth, and mixed with blurry clouds,

It immediately causes rain, and the sun shines brightly. || 30 ||

When a halo appears around the sun during the day and the moon at night.

वृष्टिवैकृताध्यायः

एकस्मिवेदहोरात्रे तदा नश्यति पार्थिवः ॥ ३१ ॥
एतेन विधिना नित्यं सप्ताहं परिविष्यते ।
सर्वभूतविनाशः स्यातास्मन्नुत्पातदर्शने ॥ ३२ ॥
त्रीणि यत्रावरुध्येरन् नक्षत्रं चन्द्रमा ग्रहः ।
त्र्यहेण वर्षन्द्र मासाद्वा जायतेभयम् ॥ ३३ ॥
सूर्यस्यविविधवर्णाः पवनेन विघट्टिताः कराः साधे ।
वियति धनुः संस्थाना ये दृप्यन्तेतदिन्द्रधनुः ॥ ३४ ॥
पश्चिमे तु दिशो भागे भवतीन्द्र धनुर्यदि ।
समेघगगनं स्निग्धं वैदूर्यविमल द्युति ॥ ३५ ॥
विद्युच्चनिर्मला भाति पूर्वे वायुर्यदा भवेत् ।
सप्तरात्रं महावर्ष निर्दिशदैवचिन्तकः ॥ ३६ |
अवृष्टौ वर्षणं कुर्यादैन्द्रीं दिशमुपाश्रितम् ।

पश्चिमायां महद्वर्षं करोतीन्द्रधनुः सदा ॥ ३७ ॥
रात्रौ चेद्दृश्यते पूर्वे भयं नरपतेर्भवेत् ।
याम्यायां बलमुख्यश्च विनाशमभिच्छाते ॥ ३८ ॥
यदान्तरिक्षे बलवान् मारुतो मारुताहतः ।
पतत्ययः स निर्घातो भवेदनिलसंभवः ॥ ३९ ॥
दिवसकृतः प्रतिसूर्यो जलकृदु दग्दक्षिणेस्थितोऽनिलकृत् ।
उभयस्थः सलिलभयं नृपमुपरि निहन्त्यधो जनहा ॥ ४० ॥

Chapter 9: The Effects of Rainfall

31. When there is rain during the day and night, the earthly troubles disappear.

32. By following this method daily, the week is protected; seeing this disturbance, the destruction of all beings takes place.

33. When the three - Nakshatra, Moon, and Planet - are obstructed together, fear arises in three days or after a month.

34. The various colors of the sun, scattered by the wind, appear in the sky; the bows that shine in the sky are Indra's bows.

35. If Indra's bow appears in the western direction, the cloud-filled sky is smooth and has the lustrous hue of a cat's eye gemstone.

36. When the clear lightning shines in the east and the wind arises, the one who contemplates divine signs predicts a great rain for seven nights.

37. In the absence of rain, perform the rain ritual by invoking the eastern direction; Indra's bow always brings great rain in the west.

38. If it is seen in the east at night, fear arises for the king; in the southern direction, the chief of the army faces destruction.

39. When the strong wind in the sky is struck by other winds, the resulting loud noise is caused by the wind.

40. The sun creates the day, water sports create fire in the south, and the wind forms both; water-based fears strike the king from above and the people from below.

उपसंहाराध्यायः १०

दुर्भिक्षमनावृष्टावतिवृष्टी शुभ्ययं परमयं च ।
रोगो नृनुभवायां नृपतिवधोऽनभ्रजातायां ॥ १ ॥
शीतोष्णविपर्यासो नो सम्यगृतुषु च संप्रवृत्तेषु ।
षण्मासाद्राष्ट्रभयं रोगभयं देवजनितं च ॥ २ ॥
अन्यत सप्ताहं प्रबंधवप्रधानन्नृपपरणं ।
रक्ते शस्त्रोद्योगो मांसास्थिवसादिभिर्मरकः ॥ ३ ॥
धान्यहिरण्यत्वक्फलकुसुमाद्यैर्वर्षितैर्भयं विन्द्यात् ।
अंगारपांशुवर्षे विनाशमायाति तन्नगरम् ॥ ४ ॥
उपला विना जलधरैर्विकृता वा प्राणिनोयदावृष्टाः ।
छिद्रं वाप्यतिदृष्टौ सस्यानां मीतिसंजननम् ॥ ६ ॥

व्यनभसीन्द्रधनुर्दिवा यदादृश्यतेऽथवा रात्रौ ।
प्राच्यामपरस्यां वा तदा भवेत्क्षुद्भयं सुमहत् ॥ ६ ॥
सूर्येन्दु पर्जन्यसमीरणानां यागःस्मृतो वृष्टिविकारकाले ।
धान्यानगो काञ्चनदक्षिणाश्च देयास्ततः शांतिमुपैतिपापं ॥ ७ ॥
अनावृष्टिकालेनरैश्चैकचितैः
गुरुं देवदेवं महेशं प्रणम्य
पयोवेत सैरुद्रपर्जन्यमूक्तैः
विधेयोऽत्रहोमो बुधैर्वर्षसिद्धये ॥ ८ ॥

Conclusion Chapter 10

In times of famine, drought, excessive rain, or other calamities,
and during illness or the killing of a king, |
there is disturbance and unrest among the people. ‖ 1 ‖

Extreme fluctuations in cold and heat, and in the proper seasons,
can cause distress for six months, along with fears of disease and
divine-induced suffering. ‖ 2 ‖

For another week, the main concern is the vulnerability of the
king and the use of weapons, causing death through blood, flesh,
bones, and the like. ‖ 3 ‖

With offerings of grains, gold, bark, fruits, flowers, and other items, one can ward off fear; but when fire and ashes rain down, destruction comes to that city. || 4 ||

When water is scarce due to the absence of rain or when creatures are deformed, there is drought, and gaps appear in the crops, causing a shortage of food. || 5 ||

When Indra's bow, the rainbow, is seen in the sky during the day or night, in the eastern or western direction, a great fear of hunger arises. || 6 ||

During times of rainfall irregularity, it is remembered that the sacrifice to the Sun, Moon, rain, and wind helps to maintain balance. Offerings of grains, cattle, gold, and gifts should be given, and thereby one attains peace and redemption from sin. || 7 ||

In times of drought, people should come together with a united mind and bow down to the great Lord Mahesh, the god of gods. By performing the Sairudra Parjanya ritual with water and conducting the prescribed fire sacrifices, wise individuals can achieve successful rainfall. || 8 ||

Epilogue

मेघोत्पत्रिगर्भाश्व गर्भस्यधारणा तथा ।
वृष्टेः शुभाशुभं चैव योगा प्राजेशसंभवाः ॥ १ ॥
स्वातियोगास्तथायोगाः सयोवर्षणदर्शकाः ।
योगाः प्रोक्ताः प्रकीर्णास्तु वृष्टिविकृति संभवाः ॥ २ ॥
वराहकश्यपादीनां, भद्रवाहोश्च संहिताम् ।
पाराशरोक्रसूत्राणि वाशिष्टीं सहितां तथा ॥ ३ ॥
समालोक्य समाकृष्य सारं तेभ्यः सुनिश्चितं ।
मयोक्ता वायुशास्त्रेऽस्मिन् योगाः प्राचीनसंमताः ॥ १ ॥
शुभे कच्छदेशे भुजंगे पुरे च
बसन्याज्ञिको वेदवेदांगवेता ।
मयाशंकरः शंकरे प्रीतियुक्तः
गुणज्ञः प्रतापी द्विवेदी प्रसिद्धः ॥ ५ ॥
प्रभाशंकराच्छास्त्रिणोलब्धबोधः
सुतः कानजित्तस्यहोरागमज्ञः ।
शकेपक्षपक्षांष्टभूसंमिते च ।
भुजे पत्तने श्रावण मास शुक्ले ॥ ६ ।
तिथौ कामदेवस्य वारेगुरोध ।
महावायुशास्त्रं समस्फुटंच ।
निमितांबुधि संतितीपोर्युपस्य ।
व्यधात्पोतरूपं बुधानंदकारि ॥ ७ ॥

समाप्तोऽयं ग्रंथः

1. Clouds, the womb of rain, and the retention of water therein,
The auspicious and inauspicious nature of rainfall, and the yogas
arising from the lord of beings.

2. Swati Yoga and other yogas, as well as the ones that indicate
rain,
Yogas are said to be diverse and scattered, arising from the
irregularities of rainfall.

3. By examining the works of Varaha, Kashyapa, and others, the
Bhadrabaho Samhita , Parashara's aphorisms, and the Vashishta
Samhita along with its commentaries,

4. Having scrutinized and extracted the essence from them, I have
confidently presented
The ancient yogas on wind science in this treatise.

5. Residing in the auspicious Kachchh region, in the city of Bhuj,
a well-versed performer of sacred rites,
By shankara, an adept in the Vedas and Vedangas, is devoted to
Shankara, a connoisseur of virtues, esteemed and valiant
Dwivedi.

6. Prabhashankara, who acquired knowledge from the scholars, is the son of Kanajitta; he is well-versed in the horary astrology.
In the Shaka era, in the Bhuj city, in the month of Shravana, when the moon is waxing,

7. On the day of Kamadeva, with Thursday as the day of the week,
The comprehensive Mahavayu Shastra has been composed, a lamp that illuminates the ocean of causes.
In this form, it dispels the darkness of ignorance.

This marks the end of the treatise.

Drona as unit of measurement of Rainfall in Ancient India

: Introduction to Drona as a unit of measurement

Drona, an ancient Indian unit of measurement, was predominantly used to measure rainfall. This unit finds its roots in the vast and diverse knowledge system of India, with its origins tracing back to Vedic and ancient Sanskrit texts. References to Drona can be found in various ancient Indian scriptures, which offer insights into the measurement system and its significance in the context of rainfall and agriculture.

: Drona in Vedic literature

The term 'Drona' appears in the Vedic literature, primarily in the Shatapatha Brahmana, a prose text that is part of the Yajurveda. The text emphasizes the importance of rituals and sacrificial ceremonies, and in this context, the Drona is introduced as a unit of measurement for grains and liquids. The Shatapatha Brahmana establishes a connection between the Drona and rainfall, as both were essential components for sustaining life and ensuring prosperity in ancient India.

: Drona in Jyotisha Vedanga

Jyotisha, one of the six Vedangas or auxiliary disciplines of Vedic studies, also provides references to the Drona as a unit of measurement. In particular, the ancient text Brihat Samhita, authored by the renowned scholar Varahamihira, offers detailed descriptions of various units of measurement, including Drona. The text discusses the measurement of rainfall in specific chapters, highlighting the importance of accurate measurement in predicting agricultural yields and determining the auspiciousness of rainfall.

: Drona in Arthashastra

The Arthashastra, an ancient Indian treatise on statecraft, economic policy, and military strategy, composed by Kautilya (also known as Chanakya), also mentions Drona as a unit of measurement. The text emphasizes the significance of measuring rainfall and monitoring water resources for the efficient management and planning of agricultural activities. In this context, the Arthashastra underscores the utility of Drona as a standard unit of measurement for rainfall, highlighting its importance in the overall administration of an empire.

: Significance and legacy of Drona

Drona's usage as a unit of measurement for rainfall in ancient India underlines the advanced understanding of the need for accurate measurement systems in agriculture and resource management. The references in Sanskrit texts, such as the Shatapatha Brahmana, Brihat Samhita, and Arthashastra, demonstrate the importance of the Drona in ancient Indian society. These texts not only provide a glimpse into the scientific knowledge of the time but also showcase the rich cultural and intellectual heritage that laid the foundation for modern measurement systems and the understanding of the significance of rainfall in sustaining life and prosperity.

Rainfall Measurements and Rainfall in Ancient India

Introduction

Rainfall and its accurate measurement have been of paramount importance throughout human history. In ancient India, a deep understanding of the significance of rainfall in sustaining life, agriculture, and prosperity led to the development of sophisticated measurement systems and techniques. This article delves into the history of rainfall measurements and the role of rainfall in ancient Indian society, drawing from various ancient Sanskrit texts and treatises.

Rainfall: The Source of Life and Prosperity

In ancient India, rainfall was considered a divine gift that played a key role in nurturing the land and ensuring its fertility. The Vedas, the oldest sacred texts of Hinduism, often mention the importance of rainfall, with the Rigveda containing hymns dedicated to the rain god, Indra, who was believed to bring forth rain to ensure bountiful harvests. This reverence for rainfall is further reflected in the various rituals, ceremonies, and prayers performed to invoke sufficient rains for a prosperous agricultural yield.

Measurement of Rainfall in Ancient India

The accurate measurement of rainfall was crucial to the efficient planning and management of agricultural activities in ancient India. Various ancient Indian texts discuss different units and methods of measuring rainfall. One such unit, the 'Drona,' finds mention in texts like the Shatapatha Brahmana, Brihat Samhita, and Arthashastra. These texts provide insights into the use of Drona and other units for measuring rainfall, emphasizing their importance in ancient Indian society.

In addition to the Drona, the Brihat Samhita by Varahamihira offers detailed descriptions of various units of measurement related to rainfall. The text discusses the significance of accurate measurement in predicting agricultural yields and determining the auspiciousness of rainfall.

Rainfall and its Impact on Agriculture and State Administration

The Arthashastra, an ancient treatise on statecraft, economic policy, and military strategy, underscores the importance of monitoring rainfall and water resources. Kautilya, the author, stresses the need for effective management and planning of agricultural activities based on accurate rainfall measurements.

The text highlights the use of the Drona as a standard unit of measurement for rainfall, showcasing its importance in the overall administration of an empire.

Moreover, ancient Indian texts often categorized rainfall as auspicious or inauspicious based on various factors, including the timing, intensity, and duration of the rain. This understanding was vital in determining the best course of action for agricultural activities, such as sowing, irrigation, and harvesting.

 Conclusion: A Rich Legacy of Rainfall Measurements in Ancient India

The attention given to rainfall measurements and the understanding of rainfall's significance in ancient India reflect the advanced scientific knowledge and intellectual heritage of the time. The references to rainfall and its measurement in ancient Sanskrit texts, such as the Vedas, Shatapatha Brahmana, Brihat Samhita, and Arthashastra, provide valuable insights into the role of rainfall in ancient Indian society.

Furthermore, these texts lay the foundation for modern measurement systems and our understanding of the importance of rainfall in sustaining life and prosperity. As we continue to grapple with the challenges of climate change and its impact on rainfall patterns, revisiting the wisdom and knowledge of ancient India can offer valuable lessons in managing our natural resources and ensuring a sustainable future for generations to come.

Layers of Atmosphere as per Vimanashastra

: Introduction to Vimanashastra and the Concept of Atmospheric Layers

The Vimanashastra, also known as the Vaimanika Shastra, is an ancient Indian text that deals with aeronautics, construction, and operation of Vimanas (aerodynamic flying machines). The text provides an elaborate and fascinating account of the Earth's atmospheric layers, which are divided into five distinct spheres: Rekha, Mandal, Kakshya, Shakti, and Kendra. These spheres together form the atmospheric region that extends up to 1600 kilometers from the Earth's surface. The Vimanashastra describes various types of turbulences and phenomena occurring within these layers, which are essential for understanding the movement of Vimanas from different realms or lokas.

: The First Layer - Rekha

The Rekha, the first and innermost layer of the atmosphere, begins at the Earth's surface, referred to as Koorna in Vimanashastra. This layer is characterized by a range of meteorological phenomena and is the primary region where Vimanas from lower realms like Bhu, Bhuvar, and Swar loka travel. The Rekha is a dense and dynamic layer that experiences various types of turbulences such as Shakti Vaat, Kiran, Shaitya, and Gharshan, which are directly influenced by the Earth's geophysical and meteorological conditions.

The Second and Third Layers - Mandal and Kakshya

The Mandal and Kakshya layers lie above the Rekha and serve as transitional zones between the Earth's surface and the higher atmospheric spheres. These layers witness the passage of Vimanas from different realms, including those from the Mahat loka. The Mandal and Kakshya layers are also subject to a variety of turbulences and phenomena similar to those present in the Rekha layer. These layers exhibit unique characteristics that are influenced by various factors such as altitude, temperature, and pressure gradients.

: The Fourth and Fifth Layers - Shakti and Kendra

The Shakti and Kendra layers are the outermost layers of the Earth's atmosphere, extending up to 1600 kilometers from the surface. The highest layer, known as Varuna, marks the boundary of the Earth's atmospheric region. The Shakti and Kendra layers are characterized by distinct atmospheric conditions and are predominantly influenced by cosmic and solar radiation. Vimanas from the Satyaloka realm do not traverse these lower atmospheric layers. These outer layers are crucial in shaping the Earth's weather patterns, climatic conditions, and are responsible for various space-related phenomena.

: The Interconnections of the Atmospheric Layers

The concept of the atmospheric layers, as described in the Vimanashastra, highlights the interconnectedness of the Earth's atmosphere and how it influences the movement of celestial vehicles, or Vimanas, from the Sapta loka. Each layer plays a unique role in shaping the Earth's environment, with specific atmospheric conditions and turbulence that affect the navigational capabilities of these ancient flying machines. The Vimanashastra offers a remarkable insight into the ancient Indian understanding of the Earth's atmospheric structure and its relationship with celestial realms. The text's detailed description of each layer and the associated phenomena reflects the profound knowledge and advanced scientific thinking of the time.

Layers	Modern names	Turbulences
Kendra	Exosphere (700 to 10000 km)	Gharshan
Shakti	Thermosphere (80 to 700 km)	Shaitya
Kakshya	Mesosphere (50 to 80 km)	Kiran
Mandal	Stratosphere (12 to 50 km)	Vaat
Rekha	Troposphere (0 to 12 km)	Shakti

© vedicvimana

Layers of Atmosphere as per Vimanashastra

Types of Vatas & Vayu layers from Mahabharata's Shanti Parva

Here from 315th chapter of Shanti Parva

भीष्म उवाच॥

नारदस्य वचः श्रुत्वा व्यासः परमधर्मवित् |
तथेत्युवाच संहृष्टो वेदाभ्यासे दृढव्रतः ॥२२॥

शुकेन सह पुत्रेण वेदाभ्यासमथाकरोत् |
स्वरेणोच्चैः स शैक्षेण लोकानापूरयन्निव ॥२३॥

तयोरभ्यसतोरेवं नानाधर्मप्रवादिनोः |
वातोऽतिमात्रं प्रववौ समुद्रानिलवेजितः ॥२४॥

ततोऽनध्याय इति तं व्यासः पुत्रमवारयत् |
शुको वारितमात्रस्तु कौतूहलसमन्वितः ॥२५॥

अपृच्छत्पितरं ब्रह्मन्कुतो वायुरभूदयम् |
आख्यातुमर्हति भवान्वायोः सर्वं विचेष्टितम् ॥२६॥

शुकस्यैतद्वचः श्रुत्वा व्यासः परमविस्मितः |
अनध्यायनिमित्तेऽस्मिन्निदं वचनमब्रवीत् ||२७||

दिव्यं ते चक्षुरुत्पन्नं स्वस्थं ते निर्मलं मनः |
तमसा रजसा चापि त्यक्तः सत्त्वे व्यवस्थितः ||२८||

आदर्शे स्वामिव छायां पश्यस्यात्मानमात्मना |
न्यस्यात्मनि स्वयं वेदान्बुद्ध्या समनुचिन्तय ||२९||

देवयानचरो विष्णोः पितृयानश्च तामसः |
द्वावेतौ प्रेत्य पन्थानौ दिवं चाधश्च गच्छतः ||३०||

Bhishma said:

22. Having heard the words of Narada, Vyasa, the knower of supreme righteousness, said 'so be it' with delight and firmly resolved in the study of the Vedas.

23. Along with his son Shuka, Vyasa then undertook the study of the Vedas. With his powerful and resonant voice, he seemed to fill the worlds.

24. As the two of them, well-versed in various forms of righteousness, continued to recite, a mighty wind blew, driven by the force of the ocean.

25. Then Vyasa, considering it a hindrance to their studies, restrained his son. Shuka, however, filled with curiosity, was undeterred.

26. O Brahman, he asked his father, "From where has this wind arisen? Please explain, for you can recount all the activities of the wind."

27. Hearing Shuka's words, Vyasa was greatly surprised and spoke the following in response to his son's query about the reason for the interruption in their studies.

28. "A divine vision has arisen in you; your mind is pure and at peace. You have cast off the darkness and passion, and are established in goodness.

29. Just as one sees one's own reflection in a mirror, you perceive your self within your self. Place your awareness in the Vedas and contemplate them with your intellect.

30. There are two paths for those who depart from this world: the path of the gods (devayana), which leads to Vishnu, and the path of the ancestors (pitriyana), which leads to the world of darkness. Both paths lead to different destinations, one upward and the other downward."

पृथिव्यामन्तरिक्षे च यत्र संवान्ति वायवः |
सप्तैते वायुमार्गा वै तान्निबोधानुपूर्वशः ||३१||

तत्र देवगणाः साध्याः समभूवन्महाबलाः |
तेषामप्यभवत्पुत्रः समानो नाम दुर्जयः ||३२||

उदानस्तस्य पुत्रोऽभूद्व्यानस्तस्याभवत्सुतः |
अपानश्च ततो ज्ञेयः प्राणश्चापि ततः परम् ||३३||

अनपत्योऽभवत्प्राणो दुर्धर्षः शत्रुतापनः |
पृथक्कर्माणि तेषां तु प्रवक्ष्यामि यथातथम् ||३४||

प्राणिनां सर्वतो वायुश्चेष्टा वर्तयते पृथक् |
प्राणनाच्चैव भूतानां प्राण इत्यभिधीयते ||३५||

प्रेरयत्यभ्रसङ्घातान्धूमजांश्चोष्मजांश्च यः |
प्रथमः प्रथमे मार्गे प्रवहो नाम सोऽनिलः ||३६||

अम्बरे स्नेहमभ्रेभ्यस्तडिद्भ्यश्चोतमद्युतिः |
आवहो नाम संवाति द्वितीयः श्वसनो नदन् ||३७||

उदयं ज्योतिषां शश्वत्सोमादीनां करोति यः |
अन्तर्देहेषु चोदानं यं वदन्ति महर्षयः ||३८||

यश्चतुभर्यः समुद्रेभ्यो वायुर्धारयते जलम् |
उद्धृत्याददते चापो जीमूतेभ्योऽम्बरेऽनिलः ||३९||

योऽद्भिः संयोज्य जीमूतान्पर्जन्याय प्रयच्छति |
उद्वहो नाम वर्षिष्ठस्तृतीयः स सदागतिः ||४०||

In the Earth and atmosphere, where the winds converge,
Seven are these paths of the wind; understand them in sequence. ||
31||

There, the powerful groups of deities known as Sadhyas came
into being,
Among them was born a son, named Samano, who was hard to
conquer. ||32||

Udana became his son, and Vyana was born as his offspring,
Next, Apana should be known, and then Prana comes after that. ||
33||

Prana had no offspring, he was invincible and a tormentor of enemies,
I shall now tell you their separate duties, just as they are. ||34||

The wind, called Prana, operates separately in all beings,
From the act of breathing in living beings, it is named Prana. ||35||

He who drives the masses of clouds and smoke and steam,
The first wind in the first path is called Pravaho Anilah. ||36||

From the clouds, the wind brings moisture, lightning, and brilliant light,
The second respiration called Avaho, moves in the sky. ||37||

He who causes the eternal rise of light from the moon and others,
The great sages call this internal movement in the body Udana. ||38||

The wind that lifts water from the four oceans,
Anilah takes up water from the Jimumutas and distributes it in the sky. ||39||

He who combines water with the Jimumutas and releases it as rain,

The third wind, called Udvaho, always moves with the best speed. ||40||

समुह्यमाना बहुधा येन नीलाः पृथग्घनाः |
वर्षमोक्षकृतारम्भास्ते भवन्ति घनाघनाः ||४१||

संहता येन चाविद्धा भवन्ति नदतां नदाः |
रक्षणार्थाय सम्भूता मेघत्वमुपयान्ति च ||४२||

योऽसौ वहति देवानां विमानानि विहायसा |
चतुर्थः संवहो नाम वायुः स गिरिमर्दनः ||४३||

येन वेगवता रुग्णा रूक्षेणारुजता रसान् |
वायुना विहता मेघा न भवन्ति बलाहकाः ||४४||

दारुणोत्पातसञ्चारो नभसः स्तनयित्नुमान् |
पञ्चमः स महावेगो विवहो नाम मारुतः ||४५||

यस्मिन्पारिप्लवे दिव्या वहन्त्यापो विहायसा |
पुण्यं चाकाशगङ्गायास्तोयं विष्टभ्य तिष्ठति ||४६||

दूरात्प्रतिहतो यस्मिन्नेकरश्मिर्दिवाकरः |

योनिरंशुसहस्रस्य येन भाति वसुन्धरा ॥४७॥

यस्मादाप्यायते सोमो निधिर्दिव्योऽमृतस्य च ।
षष्ठः परिवहो नाम स वायुर्जवतां वरः ॥४८॥

सर्वप्राणभृतां प्राणान्योऽन्तकाले निरस्यति ।
यस्य वर्त्मानुवर्तेते मृत्युवैवस्वतावुभौ ॥४९॥

सम्यगन्वीक्षतां बुद्ध्या शान्तयाध्यात्मनित्यया ।
ध्यानाभ्यासाभिरामाणां योऽमृतत्वाय कल्पते ॥५०॥

In various ways, they gather together, the dense and distinct blues,
Like clouds releasing rain, they become the thickening masses. ||
41||

By uniting, they remain undisturbed, becoming the rivers that flow,
Born for protection, they attain the nature of clouds. ||42||

The one who carries the celestial vehicles of gods in the sky,
The fourth, known as Samvaha, is the wind that crushes
mountains. ||43||

By the forceful, rough, and painful expulsion of fluids,

Clouds, when blown by the wind, do not become rain-bearing. ||44||

Roaming with fierce disturbances, thundering in the sky,
The fifth, with great speed, is named Vivaha, the wind. ||45||

In the divine flood, where waters are carried through the sky,
The sacred water of the celestial Ganges remains suspended and undisturbed. ||46||

In which the one-rayed sun is reflected from afar,
The abode of a thousand rays, where the earth shines by its light. ||47||

From which the Soma flows, the divine treasury of immortality,
The sixth, named Parivaha, is the supreme wind of swiftness. ||48||

The one who withdraws the life-breath of all living beings at the end,
On whose path, both Death and Yama follow. ||49||

Clearly observed with intellect, with a serene and eternal inner self,
By the practice of meditation, one attains immortality. ||50||

यं समासाद्य वेगेन दिशामन्तं प्रपेदिरे |
दक्षस्य दश पुत्राणां सहस्राणि प्रजापतेः ||५१||

येन सृष्टः पराभूतो यात्येव न निवर्तते |
परावहो नाम परो वायुः स दुरतिक्रमः ||५२||

एवमेतेऽदितेः पुत्रा मारुताः परमाद्भुताः |
अनारमन्तः संवान्ति सर्वगाः सर्वधारिणः ||५३||

एतत्तु महदाश्चर्यं यदयं पर्वतोत्तमः |
कम्पितः सहसा तेन वायुनाभिप्रवायता ||५४||

विष्णोर्निःश्वासवातोऽयं यदा वेगसमीरितः |
सहसोदीर्यते तात जगत्प्रव्यथते तदा ||५५||

तस्माद्ब्रह्मविदो ब्रह्म नाधीयन्तेऽतिवायति |
वायोर्वायुभयं ह्युक्तं ब्रह्म तत्पीडितं भवेत् ||५६||

एतावदुक्त्वा वचनं पराशरसुतः प्रभुः |
उक्त्वा पुत्रमधीष्वेति व्योमगङ्गामयात्तदा ||५७||

Reaching with great speed, they spread to the ends of the directions,
The ten thousand sons of Daksha, the progeny of Prajapati. ||51||

By which, once created, it does not retreat, but continues to advance,
Paravaha, the supreme wind, is named, and is difficult to overcome. ||52||

Thus, the sons of Aditi, the Maruts, are of great wonder,
Unresting, they roam everywhere, supporting all that exists. ||53||

This great wonder, that this supreme mountain,
Is suddenly shaken by the wind that blows with force. ||54||

When the exhalation of Vishnu, this wind, blows with vigor,
In a moment, the world trembles, O son. ||55||

Therefore, the knowers of Brahman do not study it excessively,
For it is said that the wind causes fear in the wind, and Brahman would be tormented. ||56||

Having spoken these words, Parashara's son, the lord,

Instructed his son, "Study it," and then went to the celestial Ganges. ||57||

References

Vayu Mahashastram Manuscript obtained from Kaiser Library , Kathmandu Nepal

Vimanashastra , Sanskrit edition published by Arya Pratinidhi Sabha

Arthashastra of Vishnugupta

Brihatsamhita of Varahamihira

Mahabharata Shanit Parva